SURVEYING MATHEMATICS MADE SIMPLE

An original book by

Jim Crume P.L.S., M.S., CFedS

Co-Authors
Cindy Crume
Bridget Crume
Troy Ray R.L.S.
Mark Sandwick L.S.I.T.

PRINTED EDITION

*****

PUBLISHED BY:
Jim Crume P.L.S., M.S., CFedS

Vertical Curves

Book 10 of this Math-Series

First publication: November, 2013

Printed by CreateSpace

Available on Kindle and other devices
Cover photo courtesy of public domain.

TERMS AND CONDITIONS

The content of the pages of this book is for your general information and use only. It is subject to change without notice.

Neither we nor any third parties provide any warranty or guarantee as to the accuracy, timeliness, performance, completeness or suitability of the information and materials found or offered in this book for any particular purpose. You acknowledge that such information and materials may contain inaccuracies or errors and we expressly exclude liability for any such inaccuracies or errors to the fullest extent permitted by law.

Your use of any information or materials in this book is entirely at your own risk, for which we shall not be liable. It shall be your own responsibility to ensure that any products, services or information available in this book meet your specific requirements.

This book may not be further reproduced or circulated in any form, including email. Any reproduction or editing by any means mechanical or electronic without the explicit written permission of Jim Crume is expressly prohibited.

Table of Contents

3

INTRODUCTION

Straight forward Step-by-Step instructions.

This book is just one part in a series of digital and printed editions on Surveying Mathematics Made Simple. The subject matter in this book will utilize the methods and formulas that are covered in the books that precede it. If you have not read the preceding books, you are encouraged to review a copy before proceeding forward with this book.

For a list of books in this series, please visit:

http://www.cc4w.net/ebooks.html

_____

Prerequisites for this book:

A basic knowledge of geometry, algebra and trigonometry is required for the explanations shown in this book.

Definitions:

Vertical Curves - Wikipedia: A parabolic curve used to connect two separate constant grade lines that creates a smooth transition from one grade line to another grade line. These include crest, sag, downhill and uphill vertical curves. There are Symmetrical Vertical Curves (equal tangents) and Asymmetrical Vertical Curves (unequal tangents).

SYMMETRICAL VERTICAL CURVE - EQUAL TANGENTS

Figure 1 shows different types of vertical curves.

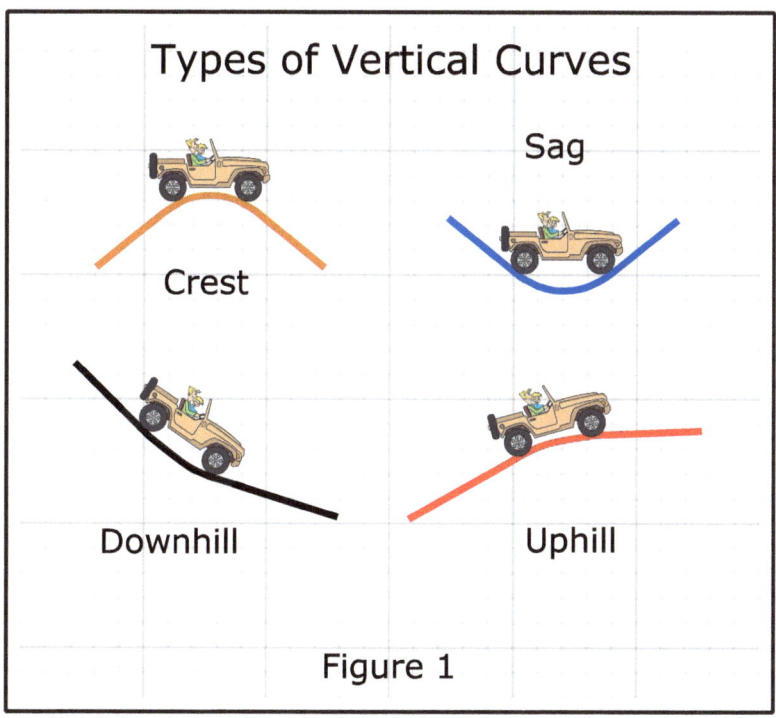

Figure 1

Figure 2 shows the various components for a vertical curve. It is important that you become familiar with these components. They will be referenced throughout this book.

For a symmetrical vertical curve the PVI is located mid point between the PVC and PVT(equal tangents).

Note: The tangent lengths are measured in the horizontal plane and NOT along the grade line. The length of the curve is measured in the horizontal plane, NOT along the curve itself.

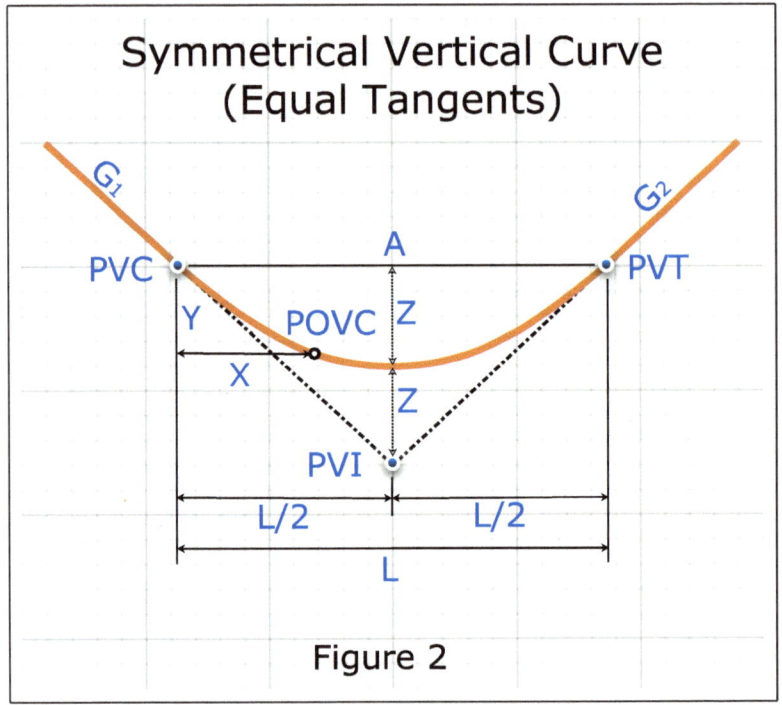

Figure 2

Figure 3 illustrates the concepts of the grade lines and the relationship with Rise over Run. Grades can be shown as percentages (i.e. 5%) or decimal equivalent (i.e. 0.05).

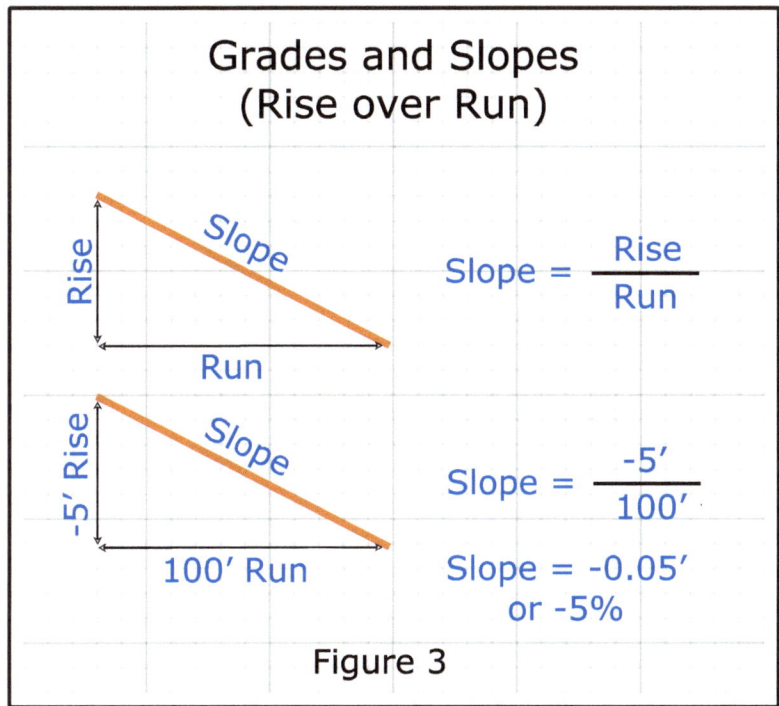

Figure 3

Definitions:

A = Mid point between PVC and PVT

G1 = Incoming Grade Line

G2 = Outgoing Grade Line

L = Vertical Curve Length in the horizontal plane

PVC = Point of Vertical Curve

PVI = Point of Vertical Intersection

PVT = Point of Vertical Tangency

POVC = Point on Vertical Curve

R = Rate of change

S = Slope of grade line

X = The horizontal distance from the PVC

Y = The elevation on the vertical curve

Z = Vertical difference between Vertical Curve to VPI and A

Formulas:

$A = (PVC_{Elev} + PVT_{Elev}) / 2$

$Z = (A + PVI_{Elev}) / 2$

$PVC_{Sta} = PVI_{Sta} - (L / 2)$

$PVT_{Sta} = PVI_{Sta} + (L / 2)$

$PVC_{Elev} = PVI_{Elev} - (G1 * (L / 2))$

$PVT_{Elev} = PVI_{Elev} + (G2 * (L / 2))$

$PVI_{Elev \, at \, POVC} = (A + PVI_{Elev}) / 2$

$R = (G2 - G1) / L$ ~ *[Rate of change]*

$Y = PVC_{Elev} + G1(X) + (R / 2)X^2$ ~ *[Elevation on curve given the X value]*

$X = (-G1 \pm \sqrt{((G1)^2 - (2 * R * (PVC_{Elev} - Y)))}) / R$ ~ *[Horizontal distance along the curve given the Y value]*

$X = -G1 / R$ ~ *[High or Low point]*

Example 1:

The following vertical curve contains a negative grade entering the curve and a positive grade exiting the curve. This is know as a Sag vertical curve.

Given:

G1 = -4.0% or -0.04

G2 = +2.0% or +0.02

8

PVISta = 10+00.00

PVIElev = 100.00

L = 400.00

POVCSta = 8+75.00

Solve for the following elements:

PVCSta = ???

PVCElev = ???

PVTSta = ???

PVTElev = ???

POVCElev = ???

PVCSta = PVISta - (L / 2)

PVCSta = 10+00.00 - (400.00 / 2)

PVCSta = **8+00.00**

PVCElev = PVI Elev - (G1 * (L / 2))

PVCElev = 100.00 - (-0.04 * (400.00 / 2))

PVCElev = **108.00**

PVTSta = PVISta + (L /2)

PVTSta = 10+00.00 + (400.00 /2)

PVTSta = **12+00.00**

PVTElev = PVIElev + (G2 * (L / 2))

PVTElev = 100.00 + (0.02 * (400.00 / 2))

PVTElev = **104.00**

R = (G2 - G1) / L

R = (0.02 - (-0.04)) / 400

R = **0.00015**

X = PVCSta - POVCSta

X = 8+00.00 - 8+75.00

X = **75.00**

POVCElev = PVCElev + G1(X) + (R / 2)X²

POVCElev = 108.00 + (-0.04 * (75.00)) + (0.00015 / 2) * 75.00²

POVCElev = **105.42**

Example 2:

Find the Low Point of the sag vertical curve.

X = -G1 / R ~ *[High or Low point]*

X = -(-0.04 / 0.00015)

X = **266.67**

LOWPOINTSta = PVCSta + 266.67 = 8+00.00 + 266.67 = **10+66.67**

Y = PVCElev + G1(X) + (R / 2)X²

Y = 108.00 + (-0.04 * (266.67)) + (0.00015 / 2) * 266.67²

Y = **102.67** ~ *[Elevation of low point]* (**See Figure 4**)

Example 3:

Find the station value for a given elevation.

Given:

POVCElev = 103.50

Solve for station:

$$X = (-G_1 \pm \sqrt{(G_1)^2 - (2 * R * (PVC_{Elev} - Y))}) / R$$

Note: There are two solutions for the quadratic equation. One or both may be a viable solution. You will need to review each solution to determine if it falls on the vertical curve.

Solution 1:

$$X = (-(-0.04) + \sqrt{(-0.04)^2 - (2 * 0.00015 * (108.00 - 103.50))}) / 0.00015$$

$X = \textbf{372.08}$

$PVC_{Sta} + 372.08 = 8+00.00 + 372.08 = \textbf{11+72.08}$

Check:

$$Y = PVC_{Elev} + G_1(X) + (R / 2)X^2$$

$$Y = 108.00 + (-0.04) * 372.08 + (0.00015 / 2) * 372.08^2$$

$Y = \textbf{103.50}$ [Checks]

Solution 2:

$$X = (-(-0.04) - \sqrt{(-0.04)^2 - (2 * 0.00015 * (108.00 - 103.50))}) / 0.00015$$

$X = \textbf{161.26}$

$PVC_{Sta} + 161.26 = 8+00.00 + 161.26 = \textbf{9+61.26}$

Check:

$$Y = PVC_{Elev} + G_1(X) + (R / 2)X^2$$

$$Y = 108.00 + (-0.04) * 161.26 + (0.00015 / 2) * 161.26^2$$

$Y = \textbf{103.50}$ [Checks]

For the Sag vertical curve, the POVC$_{Elev}$ of 103.50 is less than the PVC$_{Elev}$ (108.00) and the PVT$_{Elev}$ (104.00), therefore both solutions are viable.

Solution 1: X = 372.08 which is less than 400.00 (L) so this is a viable solution.

Solution 2: X = 161.26 which is less than 400.00 (L) so this is a viable solution.

For this vertical curve both solutions fall on the curve. **See Figure 4.**

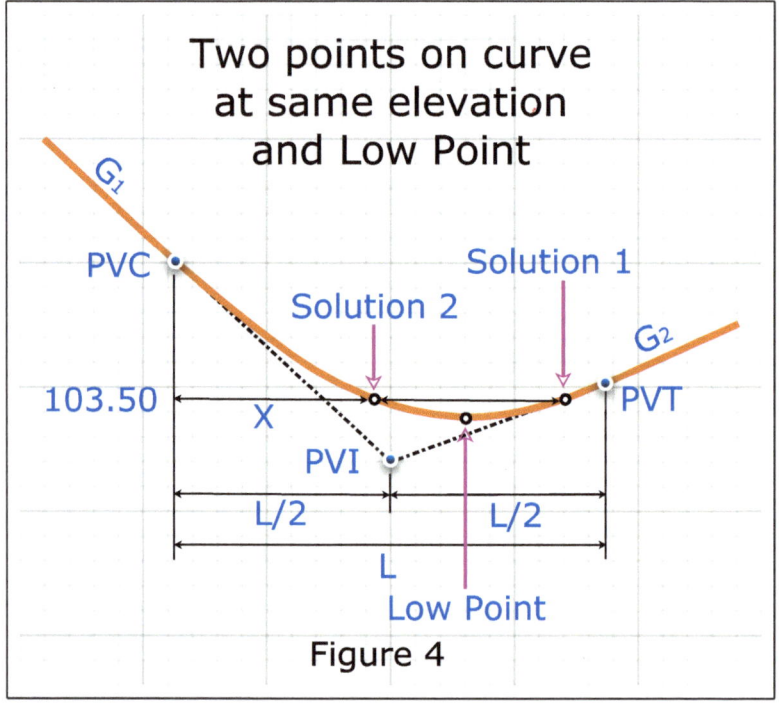

Figure 4

NOTES

ASYMMETRICAL VERTICAL CURVE - UNEQUAL TANGENTS

Figure 5 shows the various components for a asymmetrical vertical curve. It is important that you become familiar with these components. They will be referenced throughout this book.

For a asymmetrical vertical curve the PVI is <u>NOT</u> located mid point between the PVC and PVT(equal tangents).

Note: The tangent lengths are measured in the horizontal plane and NOT along the grade line. The length of the curve is measured in the horizontal plane, NOT along the curve itself.

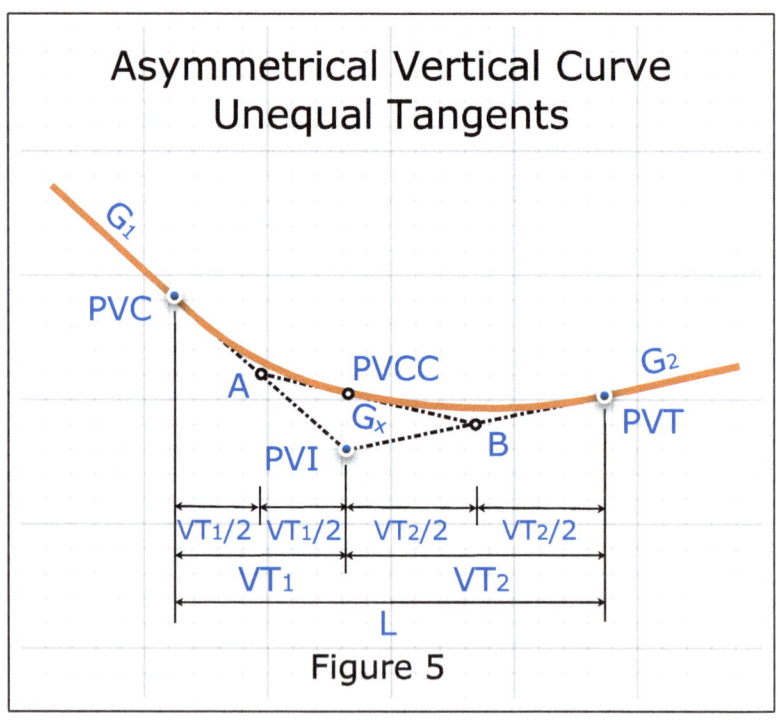

Figure 5

Additional Definitions:

A = PVI of vertical curve VT_1

B = PVI of vertical curve VT_2

G_1 = Incoming Grade Line

G_2 = Outgoing Grade Line

G_x = Grade line between A and B

PVCC = Point of Vertical Compound Curve

VT_1 = Tangent distance between PVC and PVI

VT_2 = Tangent distance between PVI and PVT

Additional Formulas:

$VT_1 = PVI_{Sta} - PVC_{Sta}$

$VT_2 = PVT_{Sta} - PVI_{Sta}$

$A_{Elev} = (PVC_{Elev} + PVI_{Elev}) / 2$

$A_{Sta} = PVC_{Sta} + (VT_1 / 2)$

$B_{Elev} = (PVI_{Elev} + PVT_{Elev}) / 2$

$B_{Sta} = PVI_{Sta} + (VT_2 / 2)$

$G_x = 2 * (B_{Elev} - A_{Elev}) / L$

$PVCC_{Elev} = A_{Elev} + (G_x * (VT_1 / 2))$

Example 4:

Solve the following asymmetrical vertical curve

Given:

G_1 = -3.0% or -0.03

G_2 = +1.0% or +0.01

PVC_{Sta} = 8+00.00

PVCElev = 206.00

PVISta = 10+00.00

PVIElev = 200.00

PVTSta = 14+00.00

PVTElev = 204.00

L = 600.00

Solve for the following elements:

VT$_1$ = ???

VT$_2$ = ???

AElev = ???

ASta = ???

BElev = ???

ASta = ???

Gx = ???

PVCCElev = ???

VT$_1$ = PVISta - PVCSta

VT$_1$ = 10+00.00 - 8+00.00

VT$_1$ = **200.00**

VT$_2$ = PVTSta - PVISta

VT$_2$ = 14+00.00 - 10+00.00

VT$_2$ = **400.00**

AElev = (PVCElev + PVIElev) / 2

AElev = (206.00 + 200.00) / 2

$A_{Elev} = \mathbf{203.00}$

$A_{Sta} = PVC_{Sta} + (VT_1 / 2)$

$A_{Sta} = 8+00.00 + (200/ 2)$

$A_{Sta} = \mathbf{9+00.00}$

$B_{Elev} = (PVI_{Elev} + PVT_{Elev}) / 2$

$B_{Elev} = (200.00 + 204.00) / 2$

$B_{Elev} = \mathbf{202.00}$

$B_{Sta} = PVI_{Sta} + (VT_2 / 2)$

$B_{Sta} = 10+00.00 + (400.00 / 2)$

$B_{Sta} = \mathbf{12+00.00}$

$G_x = 2 * (B_{Elev} - A_{Elev}) / L$

$G_x = 2 * (202.00 - 203.00) / 600.00$

$G_x = \mathbf{-0.00333 \text{ or } -0.333\%}$

$PVCC_{Elev} = A_{Elev} + (G_x * (VT_1 / 2))$

$PVCC_{Elev} = 203.00 + (-0.00333 * (200.00 / 2))$

$PVCC_{Elev} = \mathbf{202.67}$

Once you have determined the above information, use the formulas for a symmetrical vertical curve for each vertical curve segment VT_1 and VT_2 to solve for the desired components. Think of a asymmetrical vertical curve as a compound symmetrical vertical curve.

NOTES

PRACTICAL EXAMPLE

You are working on a construction project for a new roadway. The centerline profile has a symmetrical vertical curve that you need to calculate an elevation for every 50' even station and the high point for the design finish profile grade.

Given:

VPIStation = 10+00.00

VPIElev = 100.00

L = 400.00'

G1 = +2.0% or +0.02

G2 = -3.0% or -0.03

*Solve for the following elements shown in **Figure 6**:*

POVC	Station	Elevation
PVC	??	??
	??+50.00	??
	??+00.00	??
	??+50.00	??
High Pt.	??	??
PVI	10+00.00	??
	??+50.00	??
	??+00.00	??
	??+50.00	??
PVT	??	??

Figure 6

The solution can be found near the end of the book.

NOTES

SOLUTION FOR PRACTICAL EXAMPLE

Given:

VPIStation = 10+00.00

VPIElev = 100.00

L = 400.00'

G1 = +2.0% or +0.02

G2 = -3.0% or -0.03

*Solve for the following elements shown in **Figure 6:***

Start by solving for the stationing for the PVC and PVT.

PVCSta = PVISta - (L / 2)

PVCSta = 10+00.00 - (400.00 / 2)

PVCSta = **8+00.00**

PVTSta = PVISta + (L /2)

PVTSta = 10+00.00 + (400.00 /2)

PVTSta = **12+00.00**

Next solve for the elevation at the PVC, PVI and PVT.

PVCElev = PVI Elev - (G1 * (L / 2))

PVCElev = 100.00 - (0.02 * (400.00 / 2))

PVCElev = **96.00**

PVTElev = PVIElev + (G2 * (L / 2))

PVTElev = 100.00 - (-0.03 * (400.00 / 2))

$PVT_{Elev} = $ **94.00**

$A = (PVC_{Elev} + PVT_{Elev}) / 2$

$A = (96.00 + 94.00) / 2$

$A = $ **95.00**

$PVI_{Elev\ at\ POVC} = (A + PVI_{Elev}) / 2$

$PVI_{Elev\ at\ POVC} = (95.00 + 100.00) / 2$

$PVI_{Elev\ at\ POVC} = $ **97.50**

Next solve for the first +50' even station value (8+50.00).

$R = (G_2 - G_1) / L$

$R = (-0.03 - 0.02) / 400.00$

$R = $ **-0.000125**

$Y = PVC_{Elev} + G_1(X) + (R / 2)X^2$

$Y = 96.00 + 0.02 *(50.00) + (-0.000125 / 2) * 50^2$

$Y = $ **96.84**

Repeat the above process for each 50' interval by adding 50 to the X value with each iteration until you get to 350'. See **Figure 7** to check your answers.

Finally calculate the High Point for the Crest Vertical Curve.

$X = -G_1 / R$

$X = -0.02 / -0.000125$

$X = $ **160.00**

High Point Station 8+00.00 + 160.00 = **9+60.00**

$$Y = 96.00 + 0.02 *(160.00) + (-0.000125 / 2) * 160.00^2$$

$$Y = 97.60$$

Figure 7 contains the answers to the practical example.

POVC	Station	Elevation
PVC	8+00.00	96.00
	8+50.00	96.84
	9+00.00	97.38
	9+50.00	97.59
High Pt.	9+60.00	97.60
PVI	10+00.00	97.50
	10+50.00	97.09
	11+00.00	96.38
	11+50.00	95.34
PVT	12+00.00	94.00

Figure 7

NOTES

CONCLUSION

Vertical curves are not difficult to calculate when you take the divide and conquer approach. The above step by step instructions outline a simple approach to solving vertical curves that you will encounter during your surveying and engineering career.

ABOUT THE AUTHOR
Jim Crume P.L.S., M.S., CFedS

My land surveying career began several decades ago while attending Albuquerque Technical Vocational Institute in New Mexico and has traversed many states such as Alaska, Arizona, Utah and Wyoming. I am a Professional Land Surveyor in Arizona, Utah and Wyoming. I am an appointed United States Mineral Surveyor and a Bureau of Land Management (BLM) Certified Federal Surveyor. I have many years of computer programming experience related to surveying.

This book is dedicated to the many individuals that have helped shape my career. Especially my wife Cindy. She has been my biggest supporter. She has been my instrument person, accountant, advisor and my best friend. Without her, I would not be the professional I am today. Cindy, thank you very much.

Other titles by this author:

http://www.cc4w.net/ebooks.html